Sea Otter Rescue

Sea Otter Rescue

The Aftermath of an Oil Spill

Roland Smith

Photographs by the author

COBBLEHILL BOOKS
Dutton   New York

To the sea otters of Prince William Sound,
and those who helped to save them.

Acknowledgments

The author wishes to thank the following people and organizations for their help with this project: Jill and Tom Otten, Jim and Dale Styers, John and Janice Houck, Mike Jones, Jeff Zimmerman, Mel Woods, Jeff Rash, William Muñoz, Jeff Foster, The Point Defiance Zoo & Aquarium, The Valdez Convention and Visitors Center, Exxon USA, and the volunteers and staff of both the Valdez and Seward Otter Rescue centers. Special thanks goes to Dorothy Hinshaw Patent who helped me through the rough spots, and to Barbara Smith who makes all things possible—this first one's for you.

Published in the United States by Cobblehill Books, an affiliate of Dutton Children's Books, a division of Penguin Books USA Inc.
Published simultaneously in Canada by Fitzhenry & Whiteside Limited, Toronto

Designed by Charlotte Staub
Printed in Hong Kong   First Edition
10 9 8 7 6 5 4 3 2 1

Library of Congress Cataloging-in-Publication Data

Smith, Roland, date
   Sea otter rescue / Roland Smith : photographs by the author.
      p.   cm.
   Summary: Describes the rescue of the sea otters following the 1989 oil spill in Valdez, Alaska.
   ISBN 0-525-65041-5
   1. Sea otter—Alaska—Prince William Sound—Juvenile literature.
2. Wildlife rescue—Alaska—Prince William Sound—Juvenile literature. [1. Sea otter. 2. Wildlife rescue. 3. Oil spills.]
I. Title
QL737.C25S65 1990
639.9'7974447—dc20   89-49446
                                                           CIP
                                                           AC

Contents

Author's Note

A few weeks after the oil spill in Prince William Sound, I was asked to lead a team from the Point Defiance Zoo & Aquarium to help the sea otters that were being injured by the oil. When we arrived at the Otter Rescue Center (ORC) in Valdez, Alaska, we were shocked by the number of otters affected by the oil. Dozens of sea otters were being brought in every day. And volunteers from all over the United States fought courageously around the clock to save them.

After spending a day helping at the ORC, John Houck (from our team) and I were asked to go out on one of the collection boats to look for injured sea otters, while Mike Jones, the zoo's veterinarian, remained at the center to help with the efforts there.

John and I spent nearly a week on the boat searching for injured otters, during which time we saw firsthand the damage that misplaced oil wreaks on wildlife and the environment.

At night we would anchor in a sheltered bay and sleep. At daybreak we would begin searching again. When an oiled otter was spotted, it was captured and put in a crate on the boat. To save time, a helicopter or seaplane picked up the otters from the boats and flew them to the ORC.

When we got back to Valdez, because of overcrowding, arrangements were made for us to transport six of the oiled sea otters to the Point Defiance Zoo in Tacoma, Washington. We set up a facility at the zoo and spent six weeks stabilizing these animals, which were eventually put in with our resident sea otters at the zoo.

With this accomplished, I flew to Seward, Alaska, where a new ORC had been built. I spent a week in Seward photographing otters, interviewing rescuers, and helping where I could.

Again, I was struck by the volunteers' and staff's heroic efforts to save the sea otters.

The oil spill was a major tragedy that could have been avoided or, at the very least, controlled. The only positive thing about the spill is that we are now better prepared for the next time oil spills into the sea otter's home.

Prince William Sound, "Alaska's Emerald Jewel."

1

The Largest Oil Spill in the United States

Ten to fifteen thousand sea otters once lived in the cold, crystal clear waters of Alaska's Prince William Sound. The otters thrived along the sound's thousand-mile rocky shoreline.

Prince William Sound was once called "Alaska's Emerald Jewel"—until something happened that would tragically mar the beauty of the sound and destroy much of its wildlife. It all started in Valdez, a small Alaskan town on the northern edge of the sound. On the evening of March 23, 1989, a supertanker named the *Exxon Valdez* left the Valdez oil terminal and headed out through the treacherous waters of the sound, carrying within its huge dark storage hull more than 42 million gallons of thick, toxic crude oil.

The crude oil comes from the North Slope oil fields and is pumped through an 800-mile pipeline that ends at the Valdez oil terminal.

Supertankers

Despite its small size, Valdez is one of the busiest seaports in Alaska. Every day, supertankers from all over arrive at the Valdez oil terminal to fill their enormous hulls with millions of gallons of crude oil. It is estimated that over 900 tankers a year dock at the terminal for a "fill-up." The crude oil comes from the North Slope oil fields and is pumped through an 800-mile pipeline that ends in Valdez. When the crude oil arrives in Valdez it is stored in huge storage tanks until the supertankers can take it to refineries. At the refineries, the oil is made into gasoline, chemicals, plastics, and other products.

There are over 3,200 supertankers transporting oil to refineries all over the world. These tankers are called "super" because they are over 900 feet long (the size of three football fields put together) and over 150 feet wide. Some of these ships can carry over 70 million gallons of oil in one load.

Because of their size, supertankers are difficult to steer. In order to avoid hitting an obstacle, like a reef or large rock, the captain must begin to steer the huge tanker around an obstacle while the obstacle is still miles away.

The Oil Spill

When the *Exxon Valdez* left the terminal, headed for a refinery in California, the weather was unusually mild. The seas were calm, and there was very little wind. The crew was thankful for the good weather conditions, and it looked as though they would have an easy passage through the sound.

At the terminal, a harbor pilot had boarded the tanker to steer it through the hazardous waters of the harbor. Harbor pilots are hired by the state to help navigate ships safely through the shallow waters and past hidden reefs. These pilots also know how the local weather conditions and tides can affect a ship as it is leaving port.

The supertanker *Exxon Valdez* is over 900 feet long and can carry 70 million gallons of crude oil in its hull.

As the harbor pilot steered the *Exxon Valdez* through the harbor, the ship's captain stayed in his cabin located below the bridge. Shortly before they reached Rocky Point the captain came back up to the bridge. When they reached Rocky Point, the harbor pilot turned the controls over to the captain and boarded a smaller boat that took him back to the Valdez terminal.

The captain radioed the Coast Guard and told them that he was going to move the tanker from the outbound shipping lane into the inbound shipping lane to avoid small icebergs that had broken off from the Columbia Glacier about ten miles to the northwest. Shipping lanes function much like one-way streets, so that ships won't run into each other when it's dark or the weather is bad.

"The harbor pilot has disembarked at this time," the captain told the Coast Guard over the radio. "We are hooking up sea speed [a 43-minute process in which the ship builds up its cruising speed]. Estimated time of arrival [ETA] at Naked Island is zero one hundred [1 A.M.]."

"Roger that, sir," the Coast Guard said. "Request updated ice report when you get down through there."

"Okay," the captain said. "I was just about to tell you, judging from our radar, we were about to divert from the TSS [outbound shipping lane] and end up in the inbound lane if there is no conflicting traffic."

"No reported traffic," the Coast Guard said. "I've got the *Chevron California* and the *Arco Alaska* right behind them."

"That will be fine," the captain said. "We may end up over in the inbound lane. We'll notify you when we leave TSS and cross over the separation zone. Over."

"Roger that. We'll be waiting for your call. Traffic out."

Five minutes later, the captain called the traffic control center again.

"At the present time I am going to alter my course to 200 and reduce speed to twelve knots and wind my way through the ice. Naked Island ETA will be a little out of whack, but once we are clear of ice out of Columbia Bay we will give you another shout. Over."

"Roger that, sir, we'll be waiting for your call. Traffic standing by."

At 11:50 P.M. the captain turned the controls over to the third mate and told him to make a right turn back into the outbound lanes when the vessel reached a certain point near Busby Island, three miles north of Bligh Reef. At this point the captain went back down to his cabin.

The *Exxon Valdez* was now only nine minutes away from running aground on Bligh Reef.

When they reached Busby Island, the third mate ordered the helmsman to turn the tanker ten degrees to the right, just as the captain had ordered.

The rudder didn't respond. After losing one precious minute, the third mate realized that the ship was not turning. He ordered the helmsman to steer the ship another ten degrees to the right, but again the ship did not react.

A lookout ran into the ship's pilothouse to report that the flashing red

Columbia Glacier.

buoy near Bligh Reef, which should have been visible on the left side of the ship, had been spotted on the right side of the ship.

Someone in the pilothouse then realized that the automatic pilot was on and, because of this, the ship would not respond to the helmsman. By this time the tanker was only three minutes away from running aground.

The third mate immediately turned off the auto pilot and ordered the helmsman to turn the rudder full right. But by then it was too late. For three minutes they tried to get the ship to turn but, because of its size, the ship was too slow to respond.

A few minutes after midnight there was a sickening crunch of metal as the *Exxon Valdez* struck a pinnacle of rock on Bligh Reef. A few moments later the ship hit a second pinnacle. These two strikes ripped open long gashes in the hull, rupturing ten of the fifteen cargo tanks filled with crude oil.

Captains must constantly be on the lookout for icebergs that break off from Columbia Glacier.

Twelve hours went by before authorities were able to get an oil boom around the *Exxon Valdez*. By then most of the tanker's oil had spilled into Prince William Sound.

When the captain felt the first shock of the tanker hitting the rocks, he ran up to the bridge. Immediately he slowed the engines and took other measures to keep the ship from sliding off the reef, thereby preventing further damage.

In some areas of the world, tankers are required to have a "double hull." If the ship hits a reef or an iceberg there is an extra layer of steel to prevent oil from leaking out. Unfortunately, this was not a requirement in Prince William Sound. In fact, out of the 3,200 supertankers in the world, only 500 of them are equipped with double hulls.

At 12:28 A.M., about twenty minutes after striking the reef, the tanker notified the Coast Guard that it had run aground and was leaking oil.

Three hours later, personnel aboard the *Exxon Valdez* reported a loss of nearly 6 million gallons of oil, and that nothing could be done to stop it.

Special pumps called "skimmers" are used to pump the oil out of the water.

The Oil Slick

There was no oil containment equipment aboard the *Exxon Valdez,* so there was no way for the crew to stop the oil from gushing into the pure, clear waters of Prince William Sound. Twelve hours went by before authorities were able to get an oil boom around the supertanker. Oil booms float on the surface of the water and are designed to contain the oil in one spot until special pumps called "skimmers" can pump the oil out of the water into storage tanks.

Someone reported that the main mass of the oil slick was moving as if it were on a "superhighway." The slick moved at a rate of more than a mile an hour, churning the seas into a foamy oil and water mixture referred to as "chocolate mousse."

Local fishermen and their boats were hired to man oil booms. They tried to stay in front of the oil, booming off bays, streams, and valuable fish hatcheries before the oil got there. But, because of the amount of oil and the speed at which it was traveling, most of these attempts failed.

Within two days, nearly 11 million gallons of crude oil had spilled into Alaska's Emerald Jewel. During this time officials were able to do little to contain the now 32-square-mile oil slick, even though the sea was flat calm, with no wind, and the weather unseasonably warm and sunny. One of the problems was that there was not enough equipment available to handle a spill of this size.

Sixty-six hours after the *Exxon Valdez* ran aground on Bligh Reef, the balmy, calm, spring weather was suddenly replaced by a rising wind. The few oil booms that had been placed in critical parts of Prince William Sound began to blow apart in the wind, releasing the contained oil.

On Monday, March 27, winds in excess of 70 miles per hour drove the slick 35 to 40 miles into the southwest sections of Prince William Sound. All efforts to contain the oil had failed, and the slick was now considered to be out of control. It was being called the largest oil spill in United States history.

As the oil spread through Prince William Sound, the animals that lived along the shores had no idea what was coming their way. All of their lives were in jeopardy, but especially threatened was the sea otter. To understand why, we need to know something about sea otters and how they live.

Local fishermen and their boats were hired to man oil booms.

Sea otter pups are about three to four pounds at birth and grow to over thirty pounds by the end of their first year.

2

The Sea Otters of Prince William Sound

Sea otters are in the family Mustelidae, along with weasels, minks, martens, badgers, skunks, and river otters. Because the sea otter lives in salt water, scientists have classified it as a marine mammal, along with whales, dolphins, porpoises, and seals. Although the sea otter is one of the largest animals in the weasel family, it is one of the smallest of the marine mammals.

On the average, adult male sea otters weigh forty-five to ninety pounds and are about four feet long from the tips of their noses to the ends of their tails. Female sea otters are generally smaller than males, weighing forty to sixty pounds, and they are somewhat shorter. Sea otter pups are about three to four pounds at birth and grow to over thirty pounds by the end of their first year.

Sea otters have broad heads, with large dark eyes and very small ears. Their tails are thick and flat on the bottom and make up about one-third of their body length. They have large hind feet in the shape of flippers which are used to push their bodies through the water.

Unlike other marine mammals, sea otters swim mostly on the surface of the water. What's more, they usually swim on their backs, using their rear flippers to push them through the water and their tails to guide them.

The front feet are used only for gathering food and grooming, not for swimming.

At one time, the sea otter was found all around the rim of the Pacific Ocean, from Hokkaido in northern Japan, through Kamchatka Peninsula in the Soviet Union, across the Aleutian Islands of Alaska, and down the entire West Coast of North America to the Baja Peninsula.

Sea otters usually swim on their backs using their flippers to push them through the water. They use their front feet for gathering food and eating, not for swimming.

The sea otter was hunted for its soft, thick fur.

By the turn of the century many thought that the sea otter had become extinct. It was hunted mercilessly for its soft, thick fur which was used as trim for high-fashion coats. But a few sea otters managed to elude the hunters. With the help of governmental protection, the sea otter had made a remarkable comeback. There are now approximately 2,000 sea otters living off the coast of California, and perhaps as many as 100,000 sea otters living off the Alaskan shoreline.

Sea otters were very vulnerable to hunters because they can only survive in a mile-wide band along the water's edge. Sea otters need to be near shore so they can dive for their food, which lives on the sea bottom. They also use the land for shelter during bad weather, and to "haul-out" on when they get cold.

When a sea otter grooms, it starts by washing the submerged parts of its body with the palms of its front feet.

One of the differences between people and animals (especially the sea otter) is that we can make adjustments to our environment. If we get too hot we can take some clothes off, find cool water to swim in, or turn the air conditioner on. If we get too cold we can put more clothes on, find shelter, or warm ourselves by a fire. An animal like the sea otter does not have these options. Because of this, the sea otter has a slimmer margin of survival.

Staying warm when it's cold, or cool when it's hot, is called thermo-regulation. A sea otter's average body temperature is 99 to 100 degrees, which is very close to our own body temperature. Sea otters have a unique way of maintaining their body heat in frigid waters like those of Prince William Sound.

Unlike whales, porpoises, and seals, sea otters do not have a blubber layer to keep them warm in the icy water in which they live. Instead, they do something unexpected in an animal that lives in the water—they stay warm by keeping their fur dry and clean. They do this by blowing air deep into their fur where it provides an insulating layer and gives the body buoyancy in the water.

Sea otters often groom their fur, which keeps it clean. When a sea otter grooms, it starts by rubbing the submerged parts of its body with the palms of its front paws, which loosens the dirt and debris from the fur. The sea otter then does forward somersaults in the water, wetting its entire body. By using a combination of vigorous rubbing with its front paws and licking, the sea otter cleans itself from head to tail. When the sea otter is finished washing it uses its nose to blow air into the fur.

The sea otter's fur consists of two types of hair: long guard hairs, which are usually dark or pale brown, and short dense underhairs, which are silvery in color. For every guard hair there are approximately 70 underhairs, or 300,000 hairs in an area the size of a penny.

The only parts of sea otters not protected by fur are the palms of their front feet, the toe pads of their rear flippers, their ears, noses, and lips. These exposed parts make up only 1 percent of their bodies. When sea otters get cold they are able to keep all of these parts out of the cold water. This is one reason why sea otters spend so much time on their backs, with their heads, feet, and flippers in the air.

Another way that sea otters stay warm in icy waters is by eating. To keep their energy stores up, they must constantly fuel themselves with food. An otter eats one-third to one-half of its body weight every day. For a fifty-pound otter this means 5,000 to 9,000 pounds of food a year—three to four times the amount of food needed for a boy or girl of the same weight.

Sea otters eat mostly shellfish like clams, mussels, and crabs. When a

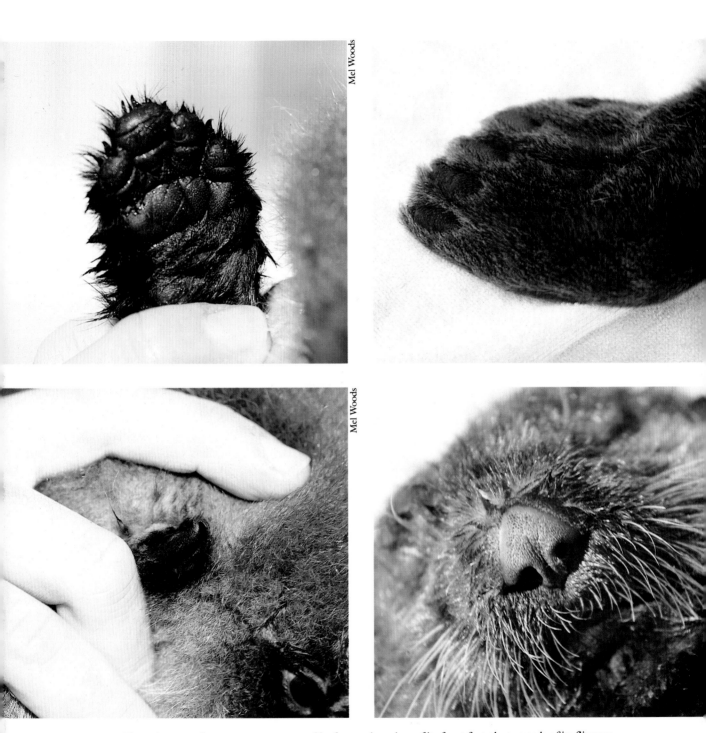

Mel Woods

Mel Woods

The only parts of a sea otter not protected by fur are the palms of its front feet, the toe pads of its flippers, its ears, nose, and lips.

An otter eats one-third to one-half of its body weight every day.

sea otter feeds, it dives to the bottom in up to thirty feet of water. It collects food with its front paws by prying mussels off rocks, or picking up crabs off the ocean floor. The sea otter tucks its prey under its "arm" for the swim back to the surface. A sea otter's lungs are quite large for the animal's size, and an otter can stay underwater for up to five minutes with one breath.

The sea otter eats its food on the surface of the water while floating on its back, using its chest as a dining table. If it cannot open the shell with its powerful jaws, it will use a rock to crack open the shell.

A baby sea otter screams in protest at being left alone.

When alarmed, sea otters will call to others nearby with high-pitched squeals, warning them of danger. When sea otters are afraid, they will often cling to each other for security. This is especially true of younger animals which, when they are scared, will cling to their mothers, or to other younger sea otters they know.

On land, sea otters are clumsy and slow, but in the water they are agile and swift. They depend on the water to elude predators. When a sea otter hauls-out on the land, it stays near the shore so it can quickly retreat into the water when it senses danger.

Breeding activities usually take place throughout the year. Pregnancy is thought to last between four and six months. Pupping peaks from January to March. A female may have her pup in the water or on land.

Baby sea otters are nearly helpless at birth, and are totally dependent on their mothers for the first few months of their lives. Besides feeding the infant, the mother constantly cleans, combs, and blow-dries the baby in order to keep it warm in the chilly water.

A baby sea otter rides on its mother's chest for the first five to eight months of its life. The only times she doesn't carry the infant are when she dives for food and when she is grooming herself.

When the baby is about a month old, the female starts to supplement the milk she is feeding her baby with solid food. At first the baby plays with the food, but eventually it starts to nibble on it. A baby sea otter is not totally weaned from its mother's milk until it is nine months to one year old.

When a mother sea otter dives for food, she leaves her pup floating on the surface by itself. She stays close to the food source so she will not have to leave her pup unattended for too long. When she returns to the surface, she checks her pup and then quickly gobbles her food down. After she has eaten, she begins grooming the pup all over, making sure that no soiling has occurred while she was away.

Despite the fact that the sea otter is now protected from hunting, its struggle for survival is far from over. Because of the characteristics of the sea otter's fur, its grooming habits, and the large amount of time that it spends on the surface of the water, the sea otter is very vulnerable to an oil spill. Direct exposure to oil causes severe soiling of the fur, which can lead to hypothermia (lowering of the body temperature). When its metabolic rate (the amount of energy it has to use in order to stay warm) increases, a sea otter will sometimes stop eating. Without food, the sea otter loses weight. Then it must expend more energy in order to stay floating on the surface of the water, since its body has lost some of the fat which gives it buoyancy. In addition to this, a sea otter may inadvertently swallow oil as it is trying to get it off its fur. Crude oil is toxic and can affect the sea otter's internal organs, like the lungs, liver, and kidneys.

Many sea otters died before the rescuers could save them.

The rocky coast of Prince William Sound is ideal habitat for sea otters.

3

When Sea Otters and Oil Collide

Ⓐll day long the sea otters had been paying close attention to the rising wind blowing into their isolated, sheltered home in a small bay in Prince William Sound. As the wind grew stronger, their nervousness increased. They knew a storm was coming, but there was no way for them to know what it would bring. By the afternoon many of them had already staked out positions in the kelp bed. Anchoring themselves to the floating tangle with their flippers, they waited patiently as the wind began to howl and the water began to swell and churn. Some of the sea otters sought refuge on land, hauling-out of the icy water onto rocks within a few feet of the shore.

A female with a month-old pup decided to risk one last dive in search of food. She was reluctant to leave her pup in the turbulent waters, but she was eating for two now and, if the storm continued to rise, she didn't know when she would get another chance to eat. Carefully, she took the sleeping pup off her chest and placed it on the surface of the water. The pup screamed loudly at such a rude awakening, but the female ignored the protest. She could see that the pup was floating nicely in the water and would be fine during her short absence.

Her powerful flippers propelled her deep into the dark silent water in search of food. Nosing her way along the side of a rock outcropping, she found a cluster of large mussels. Using her front paws, she freed two of the mussels, which she then stored in the loose folds of skin under her front legs. Having secured her catch, she pumped her flippers once, then twice, and popped through the surface. In the few seconds that she was under the water the wind seemed to have grown in intensity.

She looked for her pup and, not seeing it right away, she became alarmed. It had started to rain and the bay had become choppy, making it difficult to see very far. In the distance she heard a faint, high-pitched scream. Quickly, she swam toward the sound. After a few moments she saw her pup bobbing on the rough water.

She picked up her pup and put it on her chest, and immediately began to groom it by licking it all over and blowing air deep into its fur to dry it. When she was satisfied that the pup was dry and comfortable, she allowed the pup to nurse, and only then did she consider her own hunger.

She discovered that in her haste to find her pup, she had dropped one of the mussels. She would have to be satisfied with a partial meal. Cracking the hard shell of the remaining mussel with her teeth, she stripped out the plump orange meat. The storm was now in full force and it was time for her to find a safe spot to ride out the bad weather with her pup.

She swam over to the kelp bed and was lucky to find a reasonably sheltered area that should be safe.

The oil from the *Exxon Valdez,* pushed by the high winds and tides, found its way into the bay.

As the oil made contact with the otters, it immediately stuck to their fur. They tried to get the oil off, but the more they groomed, the deeper the oil worked its way into their fur. In the process of grooming, the sea otters swallowed large quantities of the toxic crude oil and they began to get ill.

As the oil soiled the thick fur, the icy water began to penetrate to the otters' skins. The animals began to get chilled, and lost their ability to stay afloat. The more they groomed, the more energy they burned up, and the colder they became. Some of the otters made their way to shore and hauled-out. Sadly, others, like the mother and her baby, drowned before they were able to reach the shore.

Soon the oil had spread throughout the bay. Some of the otters were badly oiled, others only slightly. But all of them were having a difficult time.

By the next morning several of the sea otters in the group were dead. The others that remained alive were struggling to stay that way.

Within twenty-four hours after the spill, animal rescue experts from all over the United States began to arrive at Valdez.

The Rescue Begins

Within twenty-four hours after the spill, animal rescue experts from all over the United States began to arrive in Valdez. Valdez is a very small town with only two hotels. Both quickly filled up with people who were trying to clean up the oil and the oiled animals. In order to accommodate the hundreds of people coming to Valdez, Exxon USA (the oil company that owned the supertanker *Exxon Valdez*) paid local residents to open up their homes, so that rescuers and clean-up crews would have places to sleep.

A scientist from Hubb's Sea World Research Laboratory was hired by Exxon to coordinate the sea otter rescue. He recruited veterinarians, zoo biologists, pathologists, toxicologists, plumbers, and carpenters from all over the country to help with the rescue.

The primary victims of past oil spills had been birds. Because very few sea otters had ever been affected by an oil spill, rescuers did not know exactly how to help the animals. At first they made do with what was available and, as they gained experience, they modified what equipment they had. They built otter washing stations out of plastic barrels cut in half

A typical sea otter washing table.

"Fish totes" were used as sea otter holding areas and swimming pools.

with screens over the tops of them, so that the oil could go down drains as the otters were being washed and rinsed. They constructed holding cages out of "fish totes," which are used by the fishing industry to transport fish to the marketplace. Plumbers piped in hot water so that rescuers could wash the otters in warm water, which helped break down the crude oil and warmed the otters up.

The first sea otter rescue center was in a community college building that had been converted into a wildlife rehabilitation center. Most of this building had to be used for the rehabilitation of oiled birds, which left only a couple of rooms available for sea otters. Because of the number of oiled otters coming through the building, it soon became clear that a separate building was needed for the injured sea otters. A large gymnasium in Valdez was rented and converted into an Otter Rescue Center.

Because Valdez was such a small town, and the nearest big city was an eight-hour drive away, getting vital supplies in the quantities that were needed was very difficult. Thousands of pounds of equipment had to be shipped in from large cities in Alaska and the Lower Forty-eight states. Among the equipment were hair driers, combs, crates, handling gloves, nets, medical drugs, rubber boots, rain suits, wire, pipe, wood, towels, veterinary supplies, hoses, nozzles—everything that was needed to set up a facility to save the sea otters.

Many of the rescuers worked twenty-four hours a day and took naps only when they could find chairs to sit in. In order to feed these people, Exxon hired a local restaurant to bring in hot meals.

By the second week of the disaster, the crude oil had spread across 3,000 square miles of Alaska's southern coastal waters. On the surface, the heavy gray goop was weathering into a sticky, tarlike substance the consistency of peanut butter. As the oil spread, pushed by the tides and winds, hundreds of sea otters were being affected.

The Seward Otter Rescue Center

As the oil got farther away from Valdez, it became increasingly more difficult to transport the injured sea otters to the rescue center in time to save them. Because of this, the decision was made to build a second sea otter rescue center in Seward, Alaska, which was closer to where the injured sea otters were being found.

Rescuers used the information they had gained from Valdez to build the Seward facility. Ten buildings were placed near the water. Each building had a separate purpose, and all of the buildings and equipment could be put in storage. If another oil spill occurred in the future, the facility could be moved by barge to the spill and reassembled.

It took workers two and a half weeks to build this facility, working twenty-four hours a day. Exxon spent over $2 million to build the Seward Otter Rescue Center, and it cost over a million dollars a month to operate it.

The Seward Otter Rescue Center was staffed with full-time employees who worked twelve-hour shifts. To assist the full-time employees, 200 to 300 volunteers a week were used. Like the Valdez Otter Rescue Center, the Seward facility used professional animal people, veterinarians, pathologists, and toxicologists from all over the United States.

Volunteers arrived daily to work for two or three weeks and, because housing was scarce, some of them had to set up tents along the beach to sleep in. These volunteers came from all walks of life—housewives, teachers, businessmen, students, lawyers, and artists. Some of the volunteers used their vacations, others simply quit their jobs and came up to Alaska to help save the sea otters.

Collecting Sea Otters

The first sea otters to come into the rescue centers were captured by containment crews working the oil booms. Eventually the rescue centers

At the Seward Otter Rescue Center ten buildings were placed near the water. Each building had a separate purpose, and all of the buildings and cages were portable, so that when the rescue effort was over they could be put in storage. The center was staffed with thirty-six full-time employees who worked twelve-hour shifts.

Collection boats were equipped with crates for transporting the injured animals.

hired local fishermen to help collect the sea otters. These boats were equipped with nets and crates for transporting the injured animals. Sometimes these crews would stay out on the water for several weeks at a time.

These collection crews had many obstacles to overcome. One problem was working in the oil-infested waters. After a day on the water, many of the crew complained of headaches and nausea. Some of this was caused by the toxic oil fumes rising off the water into the air.

Another problem they encountered was obtaining accurate information about where the oil was. As the oil slick increased in size, strong winds and tides split the slick into hundreds of smaller slicks, which would go into a bay one moment and be gone a few hours later. Therefore, it was difficult for the crews to know where oiled sea otters might be.

There were many ships, planes, and helicopters working on the oil cleanup. As much as possible they tried to stay in touch with the collection boats in order to tell them where the oil might be troubling the sea otters.

When an injured sea otter was found by clean-up crews working in the sound, they would radio one of the collection boats. The nearest collection boat would race over and pick the otter up and try to get it to the Otter Rescue Center as quickly as possible.

Many of the otters first affected by the spill were in such bad shape that collectors could literally walk up to them on the beach and put them into crates. But catching the injured otters was not always this easy. It was even sometimes difficult to tell whether an otter was oiled, especially when it was in the water.

A sea otter's security depends on staying in the water where it can maneuver freely. Only as a last resort will it haul-out on land, and, when it does, it stays on the shore so that it can get into the water as soon as it senses danger. Typically, a rescue boat would enter a bay, searching the shore with high-powered binoculars. When a suspected oiled otter was spotted hauled-out on the beach, the collection crew would get into a small inflatable boat and rush the animal, trying to cut off its access to the water. If they succeeded, the crew would jump out of the boat and net the animal. Unfortunately this did not always work. The sea otter would often get into the water before the crew was able to reach it. Since the stress associated

There were many ships working on the oil cleanup. As much as possible they tried to stay in touch with the collection boats in order to tell them where the oil might be troubling the sea otters.

with chasing it in the water could be fatal, it had to be let go. Collection crews would note where the "missed" otter was and come back in a day or two to see if they could catch it on the beach again.

Collecting sea otters could be very dangerous for the collection crew. Rescuers often fell out of their boats into the frigid waters of the sound. They wore survival suits that helped to keep them floating, but these suits did not keep the cold water out. When someone fell in the water, he or she had to be pulled out right away so as not to get hypothermia like the oiled sea otters.

Sometimes, because of bad weather, collection activities had to be suspended. At times like these, the collection boats would anchor in a sheltered bay and stay there until the storm was over.

Sea otters were sometimes caught as far as a hundred miles away from the Otter Rescue Center. To get the injured otters from the boats to the center, float planes and helicopters would meet the boats at prearranged places and times and fly the otters to the rescue center. But when the weather was bad (as it often was), the aircraft could not fly. When this happened, depending on where the collection boat was, the boat would have to make the long run into the center, which could take all day. This not only wasted valuable time for the injured sea otters, but it also took a badly needed collection boat out of commission until it could drop the otters off and get back to where other otters were being affected by the oil.

When a suspected oiled otter was spotted hauled-out on the beach, the collection crew would get into a small inflatable boat and rush the animal and try to cut off its access to the water.

Helicopters and airplanes were used to transport the oiled sea otters to the rescue center.

Several weeks after the spill, collection crews changed their techniques. Rather than rushing the sea otters when they were hauled-out on the shore, they would set tangle nets in the bays. These are nets that are strung in areas that sea otters are likely to swim through. After setting the nets, collection crews would check them every hour around the clock. When they found an otter in a net, they would untangle it and take it back to the boat. Using this technique they could catch as many as thirty sea otters in a twenty-four-hour period.

Once a sea otter was captured and taken to the center to be treated, it was still a long way from being put back into the wild.

Rescuers weigh a sea otter.

4

Caring for the Sea Otters

Many of the sea otters that were brought into the rescue center were near death, and rescuers had to struggle with many problems to save them.

Oil on the fur was not the biggest problem that had to be solved. A much more serious problem was that the otters swallowed large quantities of oil when they tried to lick it off. The crude oil was toxic and stayed inside their bodies, recirculating through their livers and kidneys, causing severe damage to these vital organs.

Surprisingly, when a mother and baby were brought in together, the baby was oftentimes relatively free of oil, but the mother was usually in very bad shape. This was because the mother spent all of her time trying to save her baby by grooming it, and in the process she virtually ignored the oil on her own fur.

The rescued sea otters were taken directly from the helicopter or airplane to the Otter Rescue Center. Once at the center, the sea otters were taken to the weighing room. After their weight was noted on a chart, the animals were taken to a room where it was quiet. There they were allowed to settle down from their traumatic capture and journey.

When a sea otter was sedated an identification tag was attached to one of its flippers. In this way rescuers could keep track of the sea otter throughout the rehabilitation process.

When it looked as though an otter had calmed down, it was moved to the sedation room. In this room, a veterinarian would inoculate it with a tranquilizing drug. Sea otters are not used to being handled by humans and, even though they were sick, they were still very strong. Without being tranquilized, they would struggle and bite, which could be very dangerous to the handlers, as well as to the sea otters themselves.

When the animal was relaxed enough, an identification tag with a number on it was clipped to one of the otter's rear flippers. This was so that volunteers and staff could keep track of the animal through the rehabilitation process. The tag could also be used to follow the sea otter's progress after it was released back into the wild.

At this point the veterinarian drew a blood sample. Blood was taken to find out if the vital organs were functioning normally.

At the Valdez center, if the otter was suspected of ingesting a lot of oil, the veterinarian would gently pass a tube through the throat to the sea

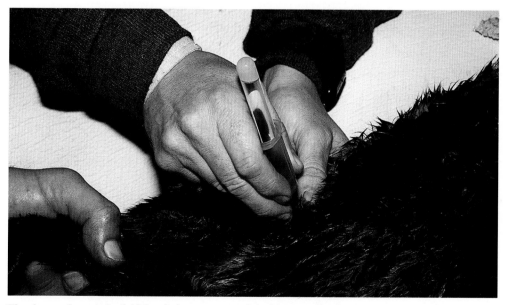

Blood was taken to check if the vital organs were functioning normally.

otter's stomach. Then a mixture of water and activated charcoal was put into the stomach through the tube. The charcoal used was very much like the charcoal in home aquarium filters and worked to "wash" the sea otter's insides. As the charcoal passed through the sea otter's digestive system, it absorbed the toxic oil and was passed out through the intestines. If this was not done, the oil would continue to circulate through the otter's body, eventually destroying the liver, which is vital to any animal's survival.

After this was done, the otter was taken to the washing room. Washing a sea otter was a very involved process that could take four people over two hours to complete, depending on how badly the otter was oiled.

In the washing room the otter was placed on a table and made as comfortable as possible by the person called the restrainer. The restrainer was always an experienced animal handler, either drawn from the staff, or a trained volunteer. The restrainer was responsible for the safety and well-being of the other washers while they worked on the otter. During the washing, if the sea otter began to wake up from the drug, the restrainer would contact the veterinarian who would then administer more tranquilizing drug to the otter. In this way the washing could continue without risk of someone being bitten.

Besides the restrainer, a minimum of three other people were needed to wash the otter—one on each side of the otter and a third at the tail end. Washing proceeded by wetting the otter with a solution of Dawn® dishwashing detergent mixed with water. The solution was poured on and then worked into the oiled fur by hand. This had to be done gently but firmly and thoroughly. Once the otter had been soaped, it was rinsed with water. Extreme care had to be taken that the rinsing water was the proper temperature. It couldn't be too cold or too warm. Rescuers not only worried about the sea otter becoming chilled, they also worried about overheating. By adjusting the temperature of the water, the sea otter's

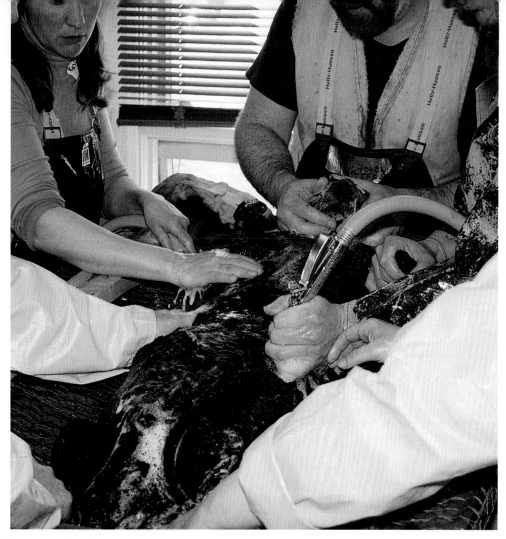

Washing and rinsing an otter could take as many as four people two hours to accomplish, depending on how badly it was oiled.

temperature could be somewhat controlled. Washing and rinsing was repeated as many as fifteen times to assure that all of the crude oil had been removed from the fur.

When the washing and rinsing cycles were complete, the otter was rinsed for an additional few minutes. It was very important that all traces of detergent be removed from the fur so that the natural oil would come back.

Once the otter had been rinsed, it was towel-dried and moved to the drying room. In the drying room, heavy duty air blowers were used to

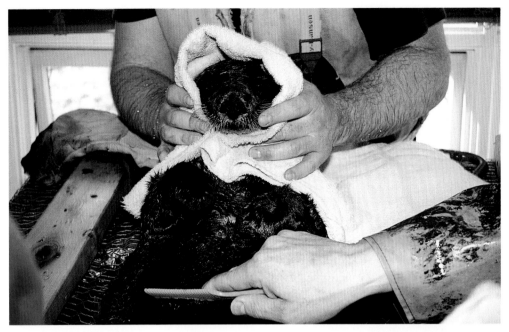

An otter being towel-dried.

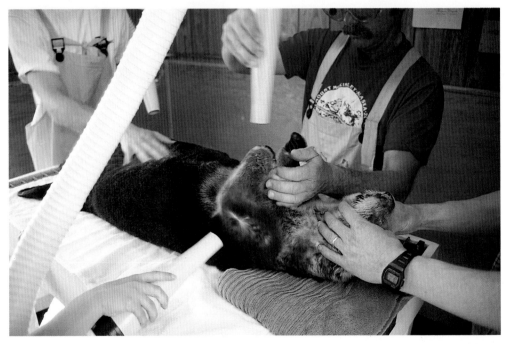

In the drying room, heavy duty air blowers were used to complete the drying process, which could take as long as an hour and a half.

complete the drying process, which could take as long as an hour and a half.

When the sea otter had been washed and dried, the veterinarian inoculated it with an "antagonist drug." An antagonist reverses the effect of the tranquilizing drug so that the otter wakes up.

From the drying room, the sea otter was moved to the recovery room. While in this room, the otter was watched closely for several hours by the veterinarian staff and treated for any problems that might arise. If it became soiled, it would be rinsed off and, if necessary, dried again. At this stage, hypothermia was still a critical problem. If the sea otter became chilled it would use all of its energy supply to get warm. If all of its energy was used up to keep warm, low blood sugar could result, which could cause the sea otter to have convulsions and go into shock. If this happened, the

A tranquilized otter being taken to the recovery unit.

veterinarian would inject dextrose under its skin, which increased the blood sugar level and helped the otter to come out of shock.

The longer a sea otter was in the Otter Rescue Center, the harder it would be to get it back into the wild. Therefore, it was important to get the animals stabilized as soon as possible.

After a few hours, if the otter appeared to be doing well in the recovery unit, it was put into the water so that it could clean and groom itself. During these swims, rescuers watched the sea otters constantly for any signs of chilling or distress. At first, the sea otter was put into a fish tote filled with water. In this way, if the otter had problems, it could be taken out of the water immediately.

Once the sea otter appeared to be doing well in the water, the length of these swims was gradually increased. When an otter was spending twenty-four hours a day in the water, it was ready to be taken to one of the bigger community tanks, where it could socialize with other otters.

If an otter appeared to be doing well in the recovery unit, it was put into the water so that it could clean and groom itself.

For these first swims, fish totes were used so that rescuers could easily get the otter out of the water at the first sign of chilling.

The sea otter's diet included shrimp, crab, clams, squid, pollock, and other fish.

Feeding the Otters

Because of the sea otter's high metabolism, feeding them the right kinds of food was very important. Food preparation was headed by a food supervisor, and staffed by one or more volunteers. The otter's food was selected by the medical staff and delivered to the food trailer for preparation.

To keep up with the hungry demands of the otters, hundreds of pounds of fresh and frozen seafood were flown into the Otter Rescue Center. This was an expensive proposition. It could cost over $1,200 a day to feed the fifty to eighty sea otters at the center.

The otters were fed five times a day, as much as they wanted to eat.

Each rescuer was responsible for a group of otters; this included feeding them.

Their diets included shrimp, scallops, crab, clams, squid, pollock, and other fish. They were also give ice blocks to chew on for fresh water and to play with.

Each rescuer was responsible for a group of otters; this included feeding them. Each otter had its own food bucket. Rescuers would bring this bucket to the food preparation workers, who would fill the bucket with the right amount and type of food. The food was put on ice to prevent it from spoiling.

In the community tanks it was necessary for two people to help with the otter feeding. As one of the rescuers fed the otters, the other rescuer would keep close watch on which otter was eating what food, and note this on their charts.

How much an otter was eating was one indication of how the otter was doing. If an otter was not eating well, it could be a sign of some kind of infection that might need treatment by the veterinarians.

One of the most important aspects of rehabilitating the sea otters was simply to watch them. By watching the sea otters twenty-four hours a day, observers got to know the individual animals very well and established "normals" for each one. In this way, if a sea otter started acting abnormally, rescuers would be able to notice it.

Observers wrote down on their charts any signs of shivering, panting, convulsions, bleeding, chronic scratching, lack of grooming, excessive screaming—anything that might be a potential problem. They also noted common behaviors, such as how much the animals were swimming, sleeping, or grooming.

Periodically the sea otters were weighed to see if they were gaining or losing weight. Because weighing was a short procedure, the otters were not

One of the most important aspects of rehabilitating sea otters was simply to watch them.

tranquilized for this. Rescuers simply scooped them out of the water with nets and carried them (net and all) over to the scale.

Mothers and Babies

Some of the female sea otters came into the rescue center with babies. Because of the mother's poor physical condition and the stress of capture, these females were usually unable to care for their babies (especially very young ones). These orphaned animals would then have to be hand-reared. Other female sea otters came in pregnant and gave birth at the Otter Rescue Center. Unfortunately, most of these had to be taken away from their mothers because the mothers were under too much stress to care for them properly.

The baby sea otters were handled very differently from the adults. A nursery was set up in one of the trailers, and a staff was specially trained to care for these babies. The staff that worked in the nursery was not allowed to have contact with any of the adult sea otters for fear that diseases would be transferred from the adults to the delicate babies.

When a sea otter is born it weighs just a little over three pounds, and it is almost totally helpless and dependent on its mother for all of its needs. Because most of the baby sea otters' nursing reflexes were not good, rescuers decided that the easiest and safest way to feed a baby was to tube the formula directly into the baby's stomach. This direct method of feeding prevented the baby from getting the formula into its lungs, which could cause pneumonia.

The babies were fed every two hours. A special formula was invented to replace the mother's milk. It was made from clams, squid, 5 percent dextrose (sugar), sterile water with minerals, half-and-half, and vitamin supplements.

When the baby sea otters were five to six weeks old, the nursery staff

The nursery staff spent a lot of time holding and grooming the babies just like their real mothers would have.

started to wean them onto solid food. To do this they would feed the babies small bits of scallops. It usually took the babies two or three days before they readily began to accept solid food. As they took in more of it, the babies were offered less of the formula until they were completely weaned.

After the babies were fed, they were taken to one of the fish totes for a swim.

After their swim, the babies were taken back to the nursery where they were towel-dried, and then blow-dried.

The nursery staff spent a lot of time holding and grooming the babies just like their real mothers would have. This nurturing was very important to the baby otters' health. When the babies were not being held or fed, they were kept on water beds at a temperature of sixty degrees. Towels were placed on the beds so that the baby otters could wrap themselves up in them if they got too cold.

After the babies were fed, they were taken to one of the fish totes for a swim. Throughout the day they were allowed in the water for three or four hours, depending on their conditions. After their swims, the babies were taken back to the nursery where they were towel-dried and then blow-dried.

Because of the close contact with their human mothers, these babies became imprinted on humans. This, unfortunately, meant that none of the babies would be able to be put back with its real mother. And therefore they could not be released back into the wild. Instead, they were sent to zoos and aquariums that had facilities to care for them properly.

Back to the Wild

When the sea otters recovered, they had to be moved out of the center in order to make room for the oiled sea otters coming in. The healthy otters were first taken to large floating pens just outside the Seward Otter Rescue Center. A floating pen has a net underneath it and on the sides, so that the otters cannot escape. Before being put into this pen they were weighed and blood was taken for the last time.

The sea otters were watched closely in this pen for a few days. If they seemed healthy, the otters were caught, inoculated with antibiotics, put into crates, and flown to floating pens that had been built in Little Jakolof Bay, near Homer, Alaska.

The pens at Little Jakolof could hold up to eighty sea otters at a time in large social groups, and human contact could be kept to a minimum. In these pens the otters were fed live food. The next step for these sea otters was freedom, but sometimes catching an animal is easier then letting it go.

Healthy sea otters were put into a large floating pen just outside of the Seward Otter Rescue Center.

A floating pen has a net underneath it and on the sides, so that the otters cannot escape.

5

Freedom

By August, four months after the oil spill, very few oiled otters were being found, and the collection boats were called in from Prince William Sound.

During the four months of the oil spill, a lot of discussion took place about the best way to set the otters free. Sea otters, like most animals, have very strong "homing instincts." In California, sea otters that had been moved to new bays were known to swim two hundred miles back to where they came from. No one knows for sure how this homing instinct works, but there is a theory that the "home bay" emits a unique chemical signal that the sea otter can detect, much in the same way a salmon can detect the stream in which it was hatched after it has spent several years in the open ocean. If the sea otters were taken directly from the rescue center and put into an unfamiliar bay, there was a chance that they would leave this bay in order to find their homes. It was feared that on their journey to their home bays they might become oiled again.

One alternative that was discussed was to set up large floating pens in clean bays that could hold groups of otters from the rescue centers. The hope was that the sea otters in these pens would set up territories, and when the pens were removed, the sea otters would stay in the bays.

But there were problems with this plan. In order to make it work, caretakers would have to stay at the pen sites to feed the otters, because the sea otters would not be able to find food in the confines of the floating pens. Another problem would be getting enough fresh food to these isolated bays. Not only would it be expensive getting the food there, but if a pen contained twenty sea otters, they would have to be fed as much as 300 pounds of food a day. Where would caretakers store all of this food to keep it from spoiling? And if the weather was bad, how would they get the food to the bays if aircraft couldn't fly?

Another alternative that was discussed was to simply leave all of the sea otters at the Otter Rescue Centers and at Little Jakolof Bay until all of the oil was out of Prince William Sound. This solution also presented some difficulties. Not only was it expensive to maintain the sea otters at the Otter Rescue Centers, but the sea otters that were in good health were only going to maintain their good health if they were set free. The conditions at the Otter Rescue Centers, while good, were not ideal to maintain a sea otter's health indefinitely.

It was finally decided to take a few of the healthy sea otters and release them with radio transmitters. In this way, biologists could follow these otters. And with the information gained from this experimental release,

Little Jakolof Bay near Homer, Alaska.

they could then determine how to proceed with the other sea otters awaiting freedom.

Radio telemetry had been used successfully for many years to study the movements of wild animals. Generally, a radio collar is placed around the neck of an animal. This collar sends out a signal to a receiver held by the biologist monitoring the animal. By determining the direction and strength of the signal, the biologist can tell where the animal is by plotting its location on a map.

In the case of the sea otters, because no collar existed that would fit their necks, biologists decided to use what is called an "implant transmitter." These transmitters work in the same way as radio collars, but instead of wearing them on their necks, they are put into the abdomens surgically. Twenty-one sea otters were implanted and released in the eastern part of Prince William Sound, which was free from oil.

Using boats and airplanes, biologists monitored the sea otters' movements for three weeks. Much to every one's delight, the biologists determined that the implanted sea otters were staying in the general area where they were being released. This meant that the other healthy sea otters could be released back into the wild.

For some of the sea otters, the successful release was considered miraculous. One of the sea otters that was released was thought to be dead when he was brought into the Otter Rescue Center three months earlier. When he arrived at the center, veterinarians could not detect a heart rate, but they continued listening. Eventually they picked up a very weak heart rate, beating only four beats a minute (a normal sea otter heartbeat is eighty

Oscar, an otter that lives in the Valdez harbor, made it through the spill without being affected at all.

beats a minute). The veterinarian staff worked on this sea otter for more than four hours, and in time they were able to revive it.

Each of the sea otters released back into the wild has its own story. And the stories will continue because of the dedication of the people who worked so hard to save the sea otters of Prince William Sound.

The Aftermath of an Oil Spill

By the time the *Exxon Valdez* oil spill was over it had traveled across over 550 miles of water, entering into hundreds of isolated coves and inlets and fouling some 800 miles of shoreline. No one knows for sure how many animals were lost, but it is estimated that well over a thousand sea otters and 32,000 birds (including over a hundred bald eagles) died as a result of the oil spill.

Of the 342 sea otters brought into the Otter Rescue Centers, nearly 75 percent were saved.

Close to a billion dollars were spent trying to clean up the oil spill, but most of the clean-up measures were not effective. The only thing that will bring Prince William Sound back to the way it was before the spill is time. And the only way to stop the damage caused by oil spills is to prevent them from happening in the first place.

Glossary

Antibiotics   Any of various substances, such as penicillin, that are used in the prevention and treatment of disease.

Buoyancy   The tendency or capacity to remain floating on the water's surface.

Convulsion   An intense involuntary muscular contraction.

Dextrose   A sugar found in animal and plant tissue derived synthetically from starch.

Groom   The act of cleaning and brushing.

Haul-out   When a seal or sea otter drag themselves out of water onto dry land.

Homing instinct   The ability to find a home.

Hypothermia   A condition of abnormally low body temperature.

Imprinting   A learning process occurring early in the life of a social animal in which the animal bonds to its (in the case of hand-reared sea otters) human mother.

Inoculate   To receive an injection of antibiotics or drugs.

Metabolism   The complex of physical and chemical processes involved in the maintenance of life.

Mustelidae   Any animal in the family Mustelidae which includes the badger, mink, otter, and weasel.

Radio collar   A device worn around the neck of mammals that sends out a signal that biologists can follow to determine where the animal is going.

Radio telemetry   The act of tracking an animal wearing a radio transmitter.

Radio transmitter   The part of a radio collar that sends out the signal.

Rehabilitation   To restore an animal to a healthy state.

Sedation   The reduction of stress or excitement by the administration of a sedative.

Tranquilizer   A drug used to calm or pacify.

Index